Re-Weaponized Prayer

Against Divisiveness

By Jimmie D. Compton, Jr.

Re-Weaponized prayer: Against divisiveness / Jimmie D. Compton, Jr.

ISBN 978-0-940123-14-4

Explores the factors contributing to the powerlessness that many Christian Westerners experience from our prayers. Also, a case is made for the reconsideration of practicing contemplative prayer. This ancient Christian form of prayer was practiced by desert solitaries beginning in the third century A.D., can help Christians today overcome several difficulties when praying.

First Edition

Dedication

Thank you, Alexandra, Barbara, Carolyn, Cedric, Chenika, Danica, Dwight, Marcus, Marquise, Nancy, Naomi, Rachel, Ray, Regina, Shelda, and Timotheus. As a result of sheltering in place during the pandemic of 2020, the death of my sister Cassandra Compton-Montgomery, and my tumor removal surgery, there were ministry responsibilities that I was unable to perform with a devoted heart. My wife Nancy made the lion's share of sacrifices for me, both personally and for our church. I thank God for her, and also that we could break away (to the porch) to celebrate forty-seven years of marriage.

Enter, your selfless hearts! Through one act of service or another, each of you made a sacrifice that brought me relief. You afforded me the peace of mind and the time to mourn, heal, recover, celebrate, and meditate in the Spirit. I am blessed to have you all in my life. This book is a product of the space you have afforded me through your selfless support.

Perspective

We worshiped Jesus instead of following him. We made Jesus into a mere religion instead of a journey toward union with God. This shift made us into a religion of belonging instead of a religion of transformation. ... One of the most subtle ways to avoid imitating someone is to put them on a pedestal, above and apart from us.

– Richard Rohr

Table of Contents

Chapter 1: A Case for Weaponized Prayer

The church in America is becoming increasingly polluted, which hinders its effectiveness as the spiritual arms and legs of Christ. How so? Her leaders have supported beliefs that do not align with biblical ethics, and in many instances, do not align with the teachings in the New Testament. Sadly, it is the most bothersome, coercive, or wealthy leaders who getting their demands met and prevail, rather than those leaders with "eyes that see and ears that hear". Unfortunately, the means by which those prevailing leaders succeed, and the consequences of their success have far-reaching effect on the community(ies) in which they reside. Often, the polarization of these two types of leaders will breed further polarization in response; either in the form of members departing or the hearts of those who remain becoming bitterly entrenched in their factional ideology. The people who suffer the most as a result of this pollution are the "least of these," who simply went into a church from sin-sick culture, seeking relief.

How can the church in America return to the focus of making disciples for Jesus!

A good start would be to first know the fundamental reasons why division will exist within a church (at any level). This goes beyond an understanding of the ideological differences on the surface. Behind the scene is a spiritual battle for influence over the human soul, willfully or by manipulation. This battle is aptly described in Job 2:1-5, and why the church in America needs Christian leaders with eyes that see and ears that hear.

The hope is, that the balance of this chapter will provide a clear contextual lens – about this heavenly battle – through which the reader can view the contents in the rest of the book.

Battle Between God and Satan
What's God Like?

God's is good (Matthew 19:17). He seeks to work in our circumstances to bring about what's good for us. *"And we know that in all things God works for the good of those who love him, who have been called according to his purpose."* Romans 8:28

Early and modern church history reveals this noteworthy fact; whenever followers of Jesus clashed with the governmental system in which they resided, those believers could not be domesticated. This is true for three reasons. Not because they had the military genius of King David, or the strength of Samson, or because they had the courage of the Maccabees. But because, 1) their faith is rooted in eternity, 2) their loyalty to Jesus precluded them us from declaring the leader of any State as supreme, and 3) God worked for their good.

Romans 8:28 is very encouraging because life can make us feel as though we are merely existing, meandering from one crisis to another without anyone who cares. But to realize that, unlike us, God actively moves for our purpose is refreshing! From the heavenly realm, He is allowing us to be pushed, pulled, sat down, sent, received and rejected on earth, in order to get us exactly where we need to be.

What's the Devil Like?

The devil's nature is evil. Evil needs no reason to hate, kill, steal or destroy. These are simply functions of evil. Christians today have been warned, "*Be alert and of sober mind. Your enemy the devil prowls around like a roaring lion looking for someone to devour.*" 1 Peter 5:8

I can testify about the importance of heeding this warning, and about the subtlety of the devil's craftiness (supported by my ignorance).

As a young Christian, I thought it was as simple as choosing right over wrong, and good over evil. Well, it didn't take too long for me to learn the hard way what I was really getting myself in to. Satan had convinced one-third of heaven's angels to follow him, and was able to deceive Eve, and his messiahs and prophets could perform great signs and wonders. Then, I had willfully and unknowingly embraced the devil's lure to sin, all while standing in the face of God, as a young minister.

Satan knows that Jesus is a more powerful opponent. He cannot beat Jesus "mano y mano"! So, he did what any political strategist worth his salt would do. Realizing that he cannot beat Jesus, Satan seeks to get leverage over Jesus by winning more souls. You and I are that leverage. However, Satan also knows that he cannot use us a leverage as long as we worship and call on the name of Jesus. He would have to trick us into thinking there is no need to worship or call on the name of Jesus. Then, he could influence our soul, and thereby use us for leverage against God. Here are two common ways Satan can trick us:

- Get us to do his evil, by convincing us that we are doing it for God. (John 16:2)

- Get us to use the weapons of this world when fighting spiritual battles. (2 Corinthians 10:3-4)

Offsetting the Subtleties of the Devil

What can Christians do to offset the subtleties of Satan? First of all, do not simply repeat what you hear other Christians say (not even the preacher). Godly preachers would want you to study the bible for yourself (2 Timothy 2:15). Second, in your zeal to fight for God, do not let the devil provoke or co-opted you into using this world's responses as weapons. When you do, it will transform the appetite of your soul to resemble the devil's more than resembling Jesus'. Instead, use God's weaponry when fighting with Him. Third and finally, remember this

For our struggle is not against flesh and blood, but against the rulers, against the authorities, against the powers of this dark world and against the spiritual forces of evil in the heavenly realms. Ephesians 6:12

Consequences of Using Spiritual Weapons

The use of God's weapons is not designed to secure our outcomes, but are designed to support a much bigger picture – the kingdom of God (of which we are citizens). Believers are indirect beneficiaries. In Luke 14:25-35, Jesus taught that there is a cost to becoming His disciple. This has proven to be the case throughout church history. Since ancient times, Christian refusal to side with the State has made us targets of persecution. But also, Jesus goes on to say in Luke 18:28-30, that there is an abundance in rewards, in this life and the life to come, for those who follow Him.

What happened to the potency of prayer between Pentecost to today, that now requires it to be re-weaponized?

Chapter 2: Biblical Enlightenment Abandoned

During the sixteenth century, the Protestant Reformation (or the European Reformation) was well on its way in challenging the practices of the Catholic Church. One might characterize it as the Church righting itself from within. However, the reformers may have lost sight of a particular uniqueness about the Church. During the Protestant Reformation, dogma, doctrine, and exegetical criticism (textual, linguistic, and cultural) bred polarization. In their protest against the Catholic Church, reformers unintentionally emphasized an either/or (or a we versus them) religious mindset. Scarce in their doctrinal positions was the role of a genuine spiritual experience. Richard Rohr states in his book *The Naked Now*,

> "Too often, religion offers more doctrinal conclusions, more competing truth claims in the increasingly large marketplace of religious claims, but seldom does it give people a vision, process, and practices whereby they can legitimate those truth claims for themselves — by inner experience and actual practices."

The days of the Reformation were tumultuous. Reformers often had to divide truth with *machetes*. Rarely did they have the luxury to leisurely parse the truth with a *scalpel*.

Response to the Age of Enlightenment

Enters the Age of Enlightenment in the seventeenth century. This movement had spread throughout Europe was fueled by the Scientific Revolution and Francis Bacon's ideas of empiricism and inductive reasoning, which preceded it. In response, and unintentional, Church reformers moved away from the emphasis of the church's unique means of divine enlightenment – prayer and meditation – to compete with the rapidly spreading intellectual, humanistic, and philosophical emphasis within the culture.

Using much of the momentum and organized faith tenets from the Reformation, the church in the West opposed the Age of Enlightenment's left-brain answers to such basic questions as "How did life begin?" "Why does humanity exist?" "How should we live?" "What is right versus wrong living?" and "Where do we go when we die?" However, in her opposition, the Western church used the same left-brain methodology (which I'd say, also yielded left-brain results). Her response overemphasized the rational aspects of the Christian faith at the expense of her potent, transrational properties (an Augustinian concept). In doing so, the Western church neglected her own unique, divine enlightenment. This is not mysticism, though it is mysterious (truth yet to be revealed). It is the divinely revealed, hidden wisdom mentioned by Moses in Deuteronomy 29:2-4, by Isaiah in chapter 6:9-10 of his book, by Jesus in Matthew 13:9-17, and by Paul in 1 Corinthians 1:17-2:16. Through either advanced theological academics (in Europe) or narrow fundamentalism (in America), the church grew lopsided, favoring the rational over the spiritual. Her ability to "see and hear" beyond the polarizing effects of human reasoning had waned. No longer was the church capable of offering her surrounding communities a more inclusive perspective to its problems and needs. Church history attests to the very polarizing perspectives she offered the souls under her influence. Here are a few:

- Catholic/Protestant
- Sinner/Saint
- Luther/Calvin
- Baptizer/Re-baptizer
- Transubstantiation/Consubstantiation

Eventually, the church lost her "What is heaven saying?" voice in the world. Such overemphasizing the rational properties of theology contributed to the emergence of denominationalism. Like peas in a single pod, followers of Christ in the West were identified by the kind of *pea* (denomination) they were in the *pod* of Christianity. They belonged to a pea that held a particular set of beliefs in common. Sadly, denominationalism seemed to be more appealing than actually experiencing the transformative power of the theology that they professed to believe rationally. Rather than *experiencing* Immanuel (God with us – Isaiah 7:14; Matthew 1:23), believers spent more time *talking about* Immanuel. Never mind that Jesus desires that we follow Him in order to be transformed through relationship and fellowship.

The mere belief-ism of the Western church is contrary to the mind of Christ. Its belief-based worship of Jesus' deity was a subtle way to avoid actually imitating Jesus' humanity in daily living. After all, right believing requires far less sacrifice than having to die-to-self daily. Also, this belief-ism makes for very shallow and powerless prayers, and it demotes the practice of prayer. Even in our prayer life today, far too many of us (self-included) often either fall asleep while praying, or mostly pray from our heads rather than through having been spiritually immersed in Christ. Mere right believing has also deluded churchgoers into *thinking* (left brain again) that we are saved, rather than *knowing* it through an experiential relationship with God.

The Western church's religious, rational response to the Age of Enlightenment's secular rationalism has proven to be one nail in its own transrational coffin. Although there is much more to God, we will never experience it living inside the belly of the whale of rationalism. We must somehow get vomited out.

Ancient Christianity's Transrationalism

The great African theologian and rhetorician Augustine of Hippo is not only considered the first modern man, but he also established the first urban monastery. He expressed the value of transrational contemplation in his book *Confessions*. In the book he credits his own salvation to the great Egyptian contemplator of antiquity, Anthony the Great. Augustine understood the value of being rational, as well as being transrational. But there were other great contemplators of Egypt in addition to Anthony the Great. In the second century A.D. there was Origen, and in the third and fourth centuries, there were Paul of Thebes, Pachomius of Thebaid, and the Theban Desert Fathers and their desert communities. These contemplators mastered the practice of transrational knowing and seeing through prayer (do not confuse this with mystics, metaphysics or manifestation). The perspective that these contemplators acquired through divine enlightenment enabled them to hold the faith's paradoxes, contradictions, incompatibilities, and mysteries in a peaceful balance with its clear and rational theological teachings. Many scholars, students, and others sought their counsel for balancing the tension between the known and the yet-to-be revealed.

Rather than fuel divine enlightenment, the either/or, right/wrong, or us/them mindset, often restricted it. The great Christian contemplators of Egypt dealt with both ambiguities and clarities through nonpolar waiting – that is allowing the space and time for divine enlightenment to emerge. They did not rush to an opinion or even speculate. The Western church has them to thank for setting the pattern for Benedict of Nursia's sixth-century introduction of monasticism to Europe.

Unfortunately, the Western church's appetite for certitude compelled it to fill the gaps left by ambiguity with academic criticism, postulates, and speculation, thereby almost abandoning contemplation altogether. A strong case can be made that this was the beginning of the decline in the perceived relevance of Christianity in the West. Do not misunderstand what I am saying. By no means does this change one jot or tittle of the truth in Scripture. In fact, Scripture had revealed nearly two millennials ago that such a polarized orientation and departure from spiritual discernment would continue to characterize humanity through the end of the Church Age. Jude tells us,

"But, dear friends, remember what the apostles of our Lord Jesus Christ foretold. They said to you, "In the last times there will be scoffers who will follow their own ungodly desires." These are the people who divide you, who follow mere natural instincts and do not have the Spirit." (Jude 1:17-19)

Chapter 3: More About Transrationalism

Biblical Expectation of Transrationalism

Polarized thinking did not begin with the Western church. To a much larger extent, the human tendency towards either/or, right/wrong, and us/them thinking has its origin in the Garden of Eden. Adam and Eve decided to act independently of God by choosing to eat the fruit from the forbidden tree – the tree of knowledge of good and evil (Genesis 2:17; 3:6-7). When they ate the forbidden fruit, Scripture tells us that their eyes were opened. Since they both already had visual sight, this is a reference to some other kind of realization. I suggest that this means that their soul's was opened to knowing – naked/clothed, guilt/innocence, good/evil, etc. The problem was, they lacked the omniscience to discern all of the contributing details required for making comprehensive judgments. The basis of their limited discernment could only take experience, memory, hearsay, or reasoning into consideration. That is a formula for making only subjective judgments that lead to polarization. In a very real sense, the disobedience of our progenitors of humanity cursed us with a bias towards polarized thinking.

Praises are due to Christ the Redeemer, that the believer's dead spirit is quickened to life by the Holy Spirit, who is deposited in us through our faith in Jesus Christ. Through the Spirit, we have access to the mind of Christ (John 16:12-14; 1 Corinthians 2:16). But in order to experience the full benefit of this access we must mature spiritually. Rohr describes the believer's spiritual maturity as being progressive.

- First, we are *pre-rational*. Naïve, childlike belief in the biblical facts taught, as described in Luke 18:17

- Next, we become *rational*: Whereby one is able to give a reason for their belief, based on God's work in history, personal experience, or observing nature, as explained in 2 Timothy 2:14-16; Hebrews 6:1-2; 1 Peter 3:15

- Then finally we become *transrational*: Willing to patiently wait for divine illumination about natural or supernatural paradoxes, contradictions, uncertainties, or conundrums without being defensive, oppositional, or compulsive, as Jesus told His disciples in Matthew 13:16-17, and Paul expounded on in 1 Corinthians 1:17–2:16

There is a spiritually unhealthy tendency in the west to remain in the rational. Apologetics and doctrine, when alone, can lead a believer to unknowingly put greater weight in reasoning, at the expense of the greater reality of God and His kingdom. Failure to mature from rational to transrational can result in placing one's faith in that which is rational about God, rather than in the all-encompassing God Himself. Among the many other problems that would result, is the problem of our omniscient, omnipotent and omnipresent God not fitting into even the best rational human mind. Because there is much more that is true about God, and His kingdom, than can be rationally comprehended.

Difficulties with Prayer

Many of the difficulties that pre-rational and rational believers experience will become almost nonexistent once they have mastered transrational practices. Contemplative prayer is an ancient Christian form of prayer that have been practiced by desert solitaries since the third century A.D. It can help Christians today overcome the following common causes of difficulties with praying:

- A lax self-denial that inhibits spiritual discipline.

- Inner dryness, whereby the soul has no taste for spiritual thoughts, memories, or feelings, nor an appetite for a Being superior to itself.

- Not taking prayer seriously due to an inability to understand what is happening on the "other end."

- Use of prayer to alter our circumstances. Then grow frustrated when God does not seem to move on our behalf, or does not move fast enough.

- The belief that prayer is not important because God is sovereign. Thinking that things will occur according to His will, regardless.

- A lack of humility, evidenced by telling God what to do in prayer.

- The belief that the power is in the praying or prayer, rather than the power being in God.

- Not recognizing God's goodness in our disappointments, thereby resulting in us being more vulnerable to Satan's attacks. He wants us to feel as though God does not care about us.

Apostle Paul's Expectation

This progression of maturity from pre-rational to transrational was clearly expected by Apostle Paul in his ministry to the church at Corinth. Acts 18 reports that Silas and Timothy joined him there during his second missionary journey in 49 A.D. The group stayed for a year and a half, preaching, gaining souls for Christ, and reasoning with those who had rejected the gospel (v. 11). Many of the Corinthians who heard Paul believed and were baptized, including Crispus, the leader of the synagogue (v. 8). Later, while on his third missionary journey in 55 or 56 A.D., Paul wrote them from Ephesus (1 Corinthians 16:8-9, 19). When I did the math, it was clear that six or seven years had passed from Paul's initial visit to Corinth until the date of his first letter to them. It is important to note here that Paul only wrote the letter in response to a letter he had received from the household of Chloe about the divisiveness among Christians in Corinth (1:10-17).

In his reply, Paul seemed frustrated by the idea that after six or seven years, the believers in Corinth were still worldly (as evidenced by their jealousy, quarreling, and gathering in cliques; vv. 1:12 and 3:3-4). He expected these believers to have progressed spiritually. Given their polarized state, he could not nurture their spiritual growth. Why? Because when one is polarized, they filter whether to relate, follow, obey, or support another person through their own polarized lens. Paul's expectation of spiritual maturity was not a form of Gnosticism that prided itself on some higher form of knowing. By that time, he expected that the believers in Corinth have matured spiritually much further, through surrendering to the Holy Spirit. The believers in Corinth had *started* with *not being receptive* toward another, and then degraded to going at each other! "To the contrary," Paul would tell them later in a second letter, "For no matter how many promises God has made, they are 'Yes' in Christ" (2 Corinthians 1:20) – yes to being open.

As a side note: Although an openness to being the "yes" in relationships has merit, in instances of heresy it is essential to draw the line to take the Lord's side. But per usual, like Jesus, the believer's starting position is a willingness to being open to listening to people who are different from us, and then waiting to hear from heaven about our response.

A Divisive Church at Corinth

1 Brothers and sisters, I could not address you as people who live by the Spirit but as people who are still worldly—mere infants in Christ. 2 I gave you milk, not solid food, for you were not yet ready for it. Indeed, you are still not ready. 3 You are still worldly. For since there is jealousy and quarreling among you, are you not worldly? Are you not acting like mere humans? 4 For when one says, "I follow Paul," and another, "I follow Apollos," are you not mere human beings? 5 What, after all, is Apollos? And what is Paul? Only servants, through whom you came to believe—as the Lord has assigned to each his task. 6 I planted the

seed, Apollos watered it, but God has been making it grow. (1 Corinthians 3:1-6)

The Corinthian believers harbored jealousy, argued back and forth when they believed that others were wrong, and gathered in cliques with those who saw things as they did. This jealousy, quarreling, and cliquishness all have the same thing in common: Each involves strict, polarized positioning. That is, observable opposition like today's Democrats vs. Republicans, Apple vs. Android users, and nerds vs. jocks.

- In jealousy the feeling is, "You're mine, not theirs."

- We quarrel, believing, "I'm right and you're wrong."

- Cliques gather around a particular preference, style, attitude, ideology, point of view, etc.

These were but different manifestations of divisiveness amongst believers in the city of Corinth.

Rohr's three-tiered spiritual rationality scale can be observed in Paul's written responses to the events in the Corinthian Church. In Chapter 1, he hears about the pre-rational jealousies, quarrels, and cliques, presumably over religious details and/or facts. He refers to their actions as worldly in 3:1-5. In Chapter 2, Paul describes the transrational enlightenment given to the apostles that they were to rely on while teaching the churches (2:6-16). Unfortunately, in Chapter 3, Paul describes his expectation that after six or seven years, believers in the church at Corinth should have reached rational maturity. It was essential that rational maturity form a basis for them to understand what Paul needed to teach regarding transrational enlightenment from the Holy Spirit (3:1), But he could not, because they were still pre-rational.

Apostle Paul's Dilemma

In Acts 18, Paul led several Corinthian souls to a relationship with our Lord Jesus as Savior. That was when they became pre-rational, according to Rohr's scale. However, upon hearing about their faith community from Chloe's household, it was apparent that they were yet to mature to the point of trusting rational teachings about the indwelling Holy Spirit, their newness in Christ, etc. Apparently, Paul expected them to (1) have embraced the rational teaching about their new creature nature – divine life inside of redeemed human flesh, and (2) live according to the Spirit – through faith and not by sight.

Because this rational teaching had not been socialized throughout their faith community, Paul was unable to disciple them to become beneficiaries of transrational knowing – an understanding received from Jesus as communicated through the same indwelling Spirt. So this age-old means of knowing, through which one acquires eyes that see and ears that hear, had not been socialized amongst them. Only a natural means for knowing had been socialized among these Corinthian believers – that is, what they could discern through their five senses, learn didactically, were told, or reason for themselves.

They were eager for spiritual gifts, and some scholars suggest that this church was abundantly gifted. In his first letter to them, Paul finds the need to manage their expectations regarding spiritual gifts (particularly 1 Corinthians chapter 12), because they were still worldly minded. A spiritually gifted church that was spiritually immature. Apparently, rather than demonstrating six or seven years of compounded spiritual maturity, Corinthian believers had simply repeated their pre-rational stage over again six or seven times.

Our great take-away from this account is that church services and works of outreach are not to be its chief characteristics. Making disciples who know Jesus through relationship, and abiding in the particular work(s) that God has predestined for each individual believer before the foundation of the world, are to be a church's chief characteristics (Ephesians 2:10; Matthew 7:21-23).

The Flesh Inhibits Spiritual Maturity

Polarized opposition is worldliness at its finest. Although it is what we ego-serving Westerners do very well, it is not the spirituality of the church (1 Corinthians 1:10), nor was it Christ's spirituality (1 Corinthians 1:13a). As much as we have invested in our opinions and beliefs, we will never mature, nor discern a *just in time* word from the Holy Spirit, if we start with *no* instead of yes—that is opposition, defensiveness, rejection, or taking sides. We cannot hear the Holy Spirit or be used by Him with an ego the starts with "no."

Jesus demonstrated how to serve God by starting with "yes." He could invite Himself to the home of a sinful tax collector (Luke 19:5-7), or eat and drink with vile people (Luke 5:30) simply by starting with yes. He knew that starting with no tends to entrench egos against Him, thereby causing it to constrict. An ego that resists starting with yes often does so to

33

prevent the

- Feeling of being compromised.

- Dreadful perception of selling out.

- Fear and shame that comes with being wrong.

If we are not humble, our own egos will entomb us in a mentality that inhibits spiritual growth and inhibits the kind of knowing that comes only from the Holy Spirit. Polarized oppositions amongst the believers in Corinth had stunted their spiritual growth, so that the growth that should have developed within six years never emerged. The resulting spiritual immaturity caused them to miss the broader objective of becoming Christ's servants themselves, just as Paul, Apollos, and Cephas had (1 Corinthians 3:5-6, 21-23). Today, the phrase might be that the believers in Corinth *kept it real or 100*. But also, they were not "people who lived by the Spirit" (1 Corinthians 3:1).

Chapter 4: What is Transrational Enlightenment?

<u>Distinction and Source</u>

So, what is this enlightenment that Moses, Isaiah, Jesus, and Paul talked about? First, we must realize what it is not:

- Some point on a continuum to perfection.

- A refusal to commit to either side.

- Being skeptical.

- Mystical or New Age thinking.

- Metaphysical manipulation of the supernatural.

- Seeing things from the other person's point of view.

- Gnosticism

- A mixture of Scripture mixed with other religions or philosophies.

- Thinking everything is relative.

- Apostasy (obviously).

Second, we must know its source. Just as a person receives the Holy Spirit upon hearing the divine supernatural call through the preaching of Christ and then believes it (Romans 10:17; Ephesians 1:13-14; Galatians 3:2-5), the same divine supernatural channel affords us enlightenment.

Transrational enlightenment is knowing or understanding something that was not derived from a human or earthly source. It may seem as trivial as whether or not to stop by the coffee shop before going work or not. Or, it could be as profound as feeling assured by knowledge received, despite your present circumstances, that you will rise from the dead in three days.

Similarities and Uniqueness

This does not mean that transrational enlightened believers do not have faith struggles. In all areas apart from what has been enlightened, they still must work out their faith walk. The significant difference is that they are in touch with their own inner struggle to not immediately to refute what is contrary, not try to quickly solve what's problematic, and not rush to shoot down parts of an opposing point of view. They are not caught up in the throes of these unaware. Instead, they actively restrain themselves from acting on their compulsion to assert that an opposing point of view is not possible, or does not make sense. Transrational enlightened believers first seek to gain perspective through contemplative prayer (Romans 8:26-27). Since they have also been rational, transrational believers are not susceptible to operate on blind faith. They are able to give a reason for their faith based on sound doctrine. They are well grounded when being open to listening to others (James 1:19). For them, certitude is not a condition that must be met before they trust, or before they will be "yes" towards others. In fact, their openness to be "yes" affirms the other person as a fellow child of the Creator, and respects their right to have a viewpoint. It is not necessarily a yes to the details of their expressed viewpoint. Transrational believers do not berate other people. Because in many instances, those people are simply repeating what they have heard or were taught without thinking through it themselves. Remember, in being open to eating and drinking with vile people (Matthew 11:19), Jesus could be God's "yes" to them (2 Corinthians 1:20).

We cannot simply settle for telling a newly redeemed believer to just believe in Jesus, without explaining the very rational works of redemption, atonement, propitiation, reconciliation, and regeneration. We have a responsibility to guide new believers along their inner journey to contemplation. They will need to experience this journey for themselves in order to attest to its enlightening manifestations in their own lives.

This enlightenment is received through openness and contemplative prayer, during which the redeemed person experiences presence with God, while hoping for perspective. In keeping with the point of this book, this hope is for perspective regarding the polarization in their midst. Such a prayer is not audible, nor is it a thinking prayer. As similar as it may seem, it is not the praying in the spirit that Paul refers to in Ephesians 6:18. It is more like prayers that Paul describes in Romans 8:26-27, where we are silent while our mind is blank. It will take time, patience, and practice (and some drifting off at times). I must note that having the proper attitude when approaching contemplative prayer is essential, particularly for refocusing ourselves when thoughts and images begin to flood our mind. An attitude that is willing to surrender our spirit to being subsumed into God's Spirit, though our silence, is key when endeavoring to pray. After all, the Spirit knows the mind of God (1 Corinthians 2:11).

Initially, our only task in contemplative prayer is the maintenance of a state of surrender to being subsumed, by remaining silent and keeping a blank mind. When we are able to do this repeatedly for about twenty minutes without agitation, then we can include asking the Lord for perspective about polarized matters. During this prayer, our focus is on our spirit being subsumed by the Holy Spirt, and not thinking about receiving a perspective from the Holy Spirit. Although God's Spirit is within us, He is not solely within us. We must not attempt to simply act on what we have read or told to believe about the Holy Spirit. The contemplative prayer experience involves an immersion into the presence of the Holy Spirit. The assumption is that we have already been walking with and talking to Him daily.

The perspective we are seeking (and sometimes not seeking) is already within us, through the indwelling Holy Spirit. However, we must desire it, and then with reverence through this kind of prayer, we can receive it. Receiving it is similar to Jacob's realization upon waking from his dream in Genesis 28:16. He dreamt about access between heaven and earth, then, upon rising he said, "Surely the Lord is in this place, and I was not aware of it."

Contemplation implies the existence of "not yet belief," a kind of questioning, or having been taken by the amazement of something. Openness affords us the necessary space for such spiritual curiosity, so the matter being pondered can be divinely refined. It is far more than trying to obtain certitude or the perfect answer to our *what should I* do question.

Igniting the Enlightenment

The power is not in the praying or the prayer. The power is in God. Prayer is our weapon that ignites God's work on our behalf in heavenly realms. Some might take offense to the aggressive nature of the following analogy, but every weapon has a source of ignition. A knife and spear have the thrust of the arm. A revolver uses a cocked hammer. A pistol has its firing pin. The arrow has the tautly strung bow. A pipe bomb uses a battery. A mine uses pressure from its target. A missile uses some type of proximity gauge. A bomb dropped from an airplane uses an altimeter. Contemplative prayer as a weapon (2 Corinthians 10:3-4) ignites access into divine enlightenment through a redeemed person's surrender, silence, and clear mind. The Holy Spirit never ceases to offer His enlightenment to the redeemed of Christ. Unfortunately, not every redeemed person values (or is taught about) such enlightenment enough to do what is necessary to acquire eyes that see and ears that hear.

Chapter 5: Restoring Transrational Enlightenment

<u>Heart and Humility of the Prayer</u>

Although 2 Kings 22 does not illustrate contemplative praying, I believe it does illustrate the kind of heart and humility that the person who prays for enlightenment should have. In that chapter, Hilkiah the high priest found Moses' Book of the Law hidden among the temple's treasure. He gave it to Shaphan the king's secretary who was with him. Then Shaphan read the book to Josiah the king of Judah. Upon hearing the words of the book of the Law, the king tore his robes and wept before the Lord. He then insisted that an

inquiry of the Lord be made on behalf of himself and the nation.

Hilkiah, Shaphan, and several others went to speak to the prophet Huldah. She revealed that the Lord was displeased with Israel's idolatry and would bring disaster upon them. She also revealed that because the king was responsive and humble when he heard the Law, the Lord would allow him to live and die in peace without seeing the disaster upon the people of Israel. After King Josiah heard the prophecy, he began a religious reform throughout the nation, starting in the temple. He restored worship in accordance with the Book of the Law. His heart, humility, and response are models for the person who engages in contemplative prayer.

With the genuineness of King Josiah, the Western church can regain its powerful voice through organic transrational enlightenment afforded her through contemplation. Otherwise, we will denigrate into just another social institution that reflects the values of the culture or the person(s) in charge. We can learn a little from Josiah about making radical changes. After seventy-five years of reigns by kings who did evil (either knowingly or unknowingly), King Josiah made radical changes within Israel's culture. It did not matter how awkward the change felt, who else was changing, or what would be lost by changing. To Josiah, the Lord was worthy of the discomfort. Also, notice in verses 13 and 18-19 that transrational space was made for the mind of God.

To expound in this book about what a leader might expect when restoring the practice of contemplation for divine enlightenment would be unfruitful. Obviously, that would depend on where things are within his or her faith community, and/or how receptive that community is to radical change. Later in this book, I will expound upon practices of contemplation that can be applied today to avoid jealousies, quarrels, and cliques among gathered believers.

A Raging Inner Storm

Although we may be open to contemplative prayer, there can be resistance from our ego, prejudice, and our defensiveness. We might not be open to reeling them in, because egos tend to trend towards a particular polar position on an issue (be it personal, social, political, philosophical, religious, etc.). Our ego, shaped by the experiences of our past, projects itself into our future, along with its fear-based defenses. Because of this, the ego rarely reflects an identity that is based solely on the current. So typically, when our ego speaks to us about our polar position, it is merely projecting into the future what we or someone else has experienced. It seeks to keep us safe, familiar and consistent with what has already happened in the past.

However, the reality is that our future is not merely more of our past, nor our present. To a large degree, the future can be shaped by present decisions. But our ego tends to harden present and future situations of life based on our past. That's why, like Jesus, when we start with yes, we are making provisions for hope for a better future.

Once we can reel our egos in, we are well on our way to becoming more like Jesus – the yes to all of God's promises. I do not want to mislead you. The practical applications of such openness in the "nasty here and now" will create resistance from your peers, especially religious ones. In particular, you will experience conflicts and misunderstandings with religious people, as in the case between Jesus and the Pharisees, Sadducees, and Zealots. Such conflicts got Jesus and most of the Lord's prophets murdered. You, too, might be misidentified as a compromiser, traitor, sellout, dreamer, an enabler of sin, or even the devil. It is simply because you are giving space for the Spirit of Truth to enlighten you, rather than parroting polarized talking points, or blindly toeing the line. It is sad, but true, that even religious people today are more responsive to optimism, positive thinking, motivational speeches, or cliches than they are to receiving a *right now* word from God.

Guard Your Mind from Your Ego

Being open to "yes in Christ" starts with disciplining the mind to *be slow to go there*, but rather defer to the Spirit of Truth. It affords us the discipline to:

- Sustain openness

- Reel in the ego's appetite for judging in order to categorize.

- Facilitate our reserving judgment.

- Leave room for appropriate listening. After all, we have two ears and one mouth, and should use them in that proportion. (James 1:19b)

On the other hand, starting at a polar end make us more susceptible to premature judgments, uninformed categorization, and mislabeling. These tends to incite our ego into defensiveness, in order to justify its position. Also, the ego must be reeled in because it only allows viewpoints that it perceives to be aligned with it. This *allowing* is not based upon what is actually true, but on the ego's perception of truth. Such a person will unknowingly try to fix what might not be broken. It just may be that it is working properly, but simply does not fit *their perception* of working properly. That is why the less judgmental our egos are, the less likely we are to only see what we desire to see.

Reeling in the ego's appetite does not mean ignoring the value of analysis, judgment, categorization, or labeling. Those are essential cerebral properties. Contemplative prayer equips us with the ability to compartmentalize the use of those properties for an appropriate context. When we have mastered the ability to reel in the ego and to compartmentalize, we will find ourselves doing so unaware. Jesus was a master at it. For instance, in the account of Him meeting the woman at the well in John 4, she brought up two very polarizing issues of the culture: ethnicity and gender (v. 9). However, Jesus did not take the bait by weighing in on either topic. Instead, He gave a non-polarized reply (v. 10). He kept the conversation on what they both had in common – being thirsty. It was the common ground He used it as a portal into her soul. Jesus' ego could have expounded on His viewpoints on the ethnic and gender issues of the day, His fairness in dealing with those issues, or which mountain was the right one for the ethnic groups to worship on. But apparently, being right about His viewpoints was of no eternal benefit. Jesus kept the focus on what His Father needed Him to be in that particular situation – an affirming and cordial person. In fact, He only mentioned her sins after she brought them up (v. 17). Jesus was not obsessed with sin management. He had more confidence in what the power of faith could do, than confidence in what could be achieved from reminding a person about their sin.

The Disarming Effect of Openness

The main character in the successful detective crime drama television series *Columbo* illustrates a particular benefit from starting with openness. Regardless of what Lieutenant Columbo thought or knew about a suspect, during an investigation he would always approach them in a disarming manner. Suspects rarely felt a need to be defensive, which compelled them to speak freely – sometimes they spoke the truth and sometimes they were deceptive. By cultivating this kind of openness, Columbo was not viewed as a threat and was often welcomed by suspects. They often felt so confident about the lies they told him, that they got sloppy.

Columbo's use of active listening along with visible gestures of approval, was followed up with his commending the suspect for wonderful explanations or solid alibis. His trademark move was to end his questioning with the usual goodbyes and then head towards the exit. Then, here's how he would get to the truth, not by refuting their claims, nor by catching them in a lie. Before leaving the actual premises, Columbo would turn around to ask the suspect one last question, which he claimed he needed their opinion in order to help him understand something. However, that question was usually loaded and crucial to breaking the whole case. It was something, which a truthful response on the part of a suspect meant revealing guilt, and a false response could easily be disproven.

Never in an episode of Columbo did he get into an ego battle with a suspect. If the suspect had ego issues, Columbo would sheepishly and humbly concede. He used relationship, egoless vulnerability, humility, and openness in a way that motivated others to remain open. Not unlike the woman at the well with Jesus, the suspects in Columbo often entangled themselves with their own dialogue.

<u>Presence with Our Creator and Others</u>

The kingdom of God is both coming (1 Corinthians 15:50), as well as already here (within the believer, Luke 17:20-21).[1] For this reason, believers are to periodically break away from our life routines to pause and re-center ourselves on what it means to be a kingdom citizen, and what its king has afforded us. Jesus often broke away to spend time alone with the Father (Matthew 26:36-45; Mark 6:46; Luke 6:12; Luke 9:18; Luke 9:28; Luke 11:1). The resulting benefits when He resumed ministry were an ability to

- Control His flesh from demeaning or being condescending to others. As Romans 14:1 puts it, "Accept the one whose faith is weak, without quarreling over disputable matters."

- Hold His thoughts captive about a person's faults while listening to them (2 Corinthians 10:5).

- Remain actively present with the person speaking. That is, He would hone in on what a person was saying, and how they said it, rather than simply wait His turn to share His opinion. He attended to the person's present misery, rather than point their past sins. He spoke about His

Father's interest in them, rather than quote great philosophers.

- Be transparent while being present with others. This enabled Him to see the common ground and be a gifted teacher.

  o Jesus' ego did not become puffed up in a manner that would differentiate Himself from others.

  o He did not need to show off what He knew.

Divine presence through prayer, an egoless faith walk, and genuine care for the soul of others was the way to the Father that Jesus had shown His disciples (John 14:1-7). They too, were to reel in their egos and compartmentalize their mind's appetite. By doing so, they, too could develop eyes to see and ears to hear. Later, as apostles, this enlightenment would supplement their five senses, their memory of Jesus' teachings, their personal experiences, and their own reasoning.

Likewise, as we make our way to the Father, our egos will not constrict and become closed to receiving people who are different from us, or believe differently from us. We will acquire the patience to wait on the Holy Spirit's enlightenment, just as Jesus promised in Matthew 13:16-17 and John 16:13. Over time, it will be natural to remain open when relating to other people, regardless of their religious affiliation, or their lack of one. We will not allow our ego to seek clues or reasons to categorize the person, or to justify our own position. Humbly and deliberately, we will choose to reserve judgment until the Holy Spirit illuminates either, something about us or something about the other person, that would inform us about the matter at hand. From my experience, often the Spirit has done so long after I have spoken with the person. And that is exactly what He wanted – for me to say nothing at the time!

The alternative would be devastating. The ego's rush to speak from a polar position tends to wrap pride or defensiveness around our position. Should the Holy Spirit later illuminate something contrary to the position we have taken, then we are faced with having to either, quench the Spirit by moving on ignoring His conviction, or seek the person out to apologize and correct ourselves. None of those are comfortable to do. Nor would we want to cause the other person's ego to wrap defensiveness around their position, thereby inhibiting their spiritual illumination and making it undesirable for them to seek us out again. Therefore, while conversing with a person who has a different viewpoint, do not allow you ego to lock you in a cocoon of defensiveness and defiance. It will stunt your spiritual growth.

Enlightenment from contemplative prayer can help bring clarity in those instances where we tend to take a polar stance on ambiguous matters. It keeps us from needing to be right all the time, or needing all the pieces in place before getting along with others. The fruit from practicing contemplation can prevent personal failure due to polarization, bouncing from one polar end to the other, or yielding control to the ego to define who you are. This kind of prayer is not simply a different thought process, behavior, belief, or fad. It has transformative powers.

The transformation from a belief-based religiosity to a relationship-based spirituality is essential for genuine fellowship with God. It is the very means that enables us to live as Jesus commands (Luke 6:46; John 14:15). He has given us commands as well as the means to know and keep them.

---

[1] Unlike some other Bible translators, the New International Version Bible translators chose to translate the Greek term *ento* in Luke 17:21 to mean "midst." It is true that ento was used as such 400 years before Jesus' birth by Greek philosophers Xenophon and Plato. However, by the time of Jesus' birth, the term ento was used to mean "inside."

• See Matthew 23:26.

- In Egypt, 300 years after Jesus' ascension ento was used to mean within in the Oxyrhynchus Papyrus sayings of Jesus regarding Luke 17:21 (Adolf Deissmann, Light from the Ancient East, 426).

Chapter 6: There's Power in Love and in Suffering

If you have ever loved before or have suffered before, then you are uniquely equipped to pray contemplatively. The irony is that, though love and suffering are great polarizers, they are also the best facilitators for receiving enlightenment through contemplation. Both can polarize as well as depolarize. This should be no surprise because, after all, God used the vulnerability of His own love, and the suffering of His Son, to bring salvation and His Spirit to all who will believe.

Love Enables Surrender

Love often compels us to bind ourselves to a particular position or person. More often than not, surrendering impresses the one we love, or we surrender for the love and attention received when advocating a position, or the self-validation that we feel when advocating the position. Unfortunately, often, such a love has very little to do with the intrinsic nature of the position. We simply love the reward received when we promote it.

Having bound ourselves to a position, we can willingly give up control to it, and thereby become vulnerable to what such a bond may bring. That is how love facilitates contemplative prayer. Through loving, we have already practiced the task of willfully giving up control. This familiarity with surrendering has prepared us with being subsumed while contemplating. It is very much like a child's willingness to allow a loving parent to take control. The same surrendering to God occurs during contemplation.

Suffering Enables Surrender

Be it abandonment, neglect, betrayal, or marginalization, when we must endure emotional distress, even due to no fault of our own, the weight of the pain can be unbearable.

As repulsive as it may feel, we have few options but to bear it all. In response, the ego attempts to protect itself by making inner vows, which have strong polarizing effects. "I'll never love again!" "I'm going to only look out for what's best for me!" "I'll never trust _____, ever again!" "I'll never let anyone get that close to me again!" These "never again" vows render us vulnerable to polarization. Unlike the case with love, in suffering we give up control unwillingly. Nevertheless, we also practice the task of surrender (albeit as a means of survival). This relinquishing of control in suffering prepares us for surrendering and being subsumed during contemplative prayer. Letting go when suffering eventually freed Job to pray for his friends (Job 42:10), who had previously argued with him (polarization) during his time of trouble.

But loving and suffering can also obstruct contemplative prayer. A threat exists in loving when we have an unhealthy attachment to someone. I am not referring to the obvious reasons regarding abusive relationships. But the unhealthy love of cherishing someone more than our Creator. Also, a threat exists in suffering when we are more committed to our inner vows than we are committed to our Creator. These threats will inhibit our ability to effectively surrender during contemplation,

59

thereby preventing genuine enlightenment. In order to use us, God may have to allow great emotional pain in our unhealthy love relationship. In the case of unhealthy inner vows, God may have to permit grave consequences to come upon us as a result of those inner vows.

Jesus Depolarizes the Polarizers

The Pharisees tried to force Jesus to pick a side - pro-government or pro-rebellion. In His reply, He did not feel compelled to stay within their stated limitations.

"Then the chief priests, the teachers of the law and the elders looked for a way to arrest him because they knew he had spoken the parable against them. But they were afraid of the crowd; so they left him and went away. Later they sent some of the Pharisees and Herodians to Jesus to catch him in his words. They came to him and said, "Teacher, we know that you are a man of integrity. You aren't swayed by others, because you pay no attention to who they are; but you teach the way of God in accordance with the truth. Is it right to pay the imperial tax to Caesar or not? Should we pay or shouldn't we?" But Jesus knew their hypocrisy. "Why are you trying to trap me?"

he asked. "Bring me a denarius and let me look at it." They brought the coin, and he asked them, "Whose image is this? And whose inscription?" "Caesar's," they replied. Then Jesus said to them, "Give back to Caesar what is Caesar's and to God what is God's." And they were amazed at him." (Mark 12:12-17)

The chief priests sent the Pharisees because they were a religious party who had a civil agreement with Rome. They posed two polarized positions to Jesus: (1) Should we pay taxes to the leader of a foreign reign (Rome) who imposes oppressive legislation upon God's people living in and around of Jerusalem? or (2) Should we refuse to pay the reign taxes, as the Zealots who wanted to overthrow that reign? To stack the deck in their trap against Jesus, the Pharisees took the Herodians along with them to confront Jesus with the question.

The reason the chief priests sent the Herodians as well, was because they were a family-based political party who also had an agreement with the Romans. In effect, the chief priests, teachers of the law, and the elders were using polarization as a tool in an attempt to trap Jesus into taking a position in support of Rome (like the Pharisees and Herodians) or against Rome (like the Zealots). Polarization has been the source of wars, break-ups, divorces, and ruined relationships. This was no different. Jesus certainly would be technically correct had He proclaimed, *I am that I am, and hereby declare Caesar to be an intruder in my holy city of Jerusalem. Therefore, because of his contempt for my people, he does not deserve to receive tax money from them.* But that would have an ego-polarizing claim. Refusing to be drawn into their polarizing framework, Jesus disarmed them by responding with a non-polarized framework. They tried to frame the issue as Rome or God (suggesting that it cannot be both), but Jesus framed it as Rome and God (suggesting that this matter can be both).

This serves as an example of the kind of shrewd wisdom that we Christians must have while navigating through our divisive and confused culture. We must be cognizant at all times of the social pressure compelling us to pick a side of the divisive "flavor of the week." The social pressure to choose a side is enormous. That is why enlightened wisdom resulting from having eyes that see and ears that hear is essential.

Chapter 7: Contemplative Prayer Can Depolarize

The following are seven ways to shine our light in the culture using contemplative prayer:

1. Many of us have been socialized to think in a binary fashion (either/or, right/wrong, for/against, etc.). When you have mastered the practice of contemplation, almost intuitively and very often, you will receive insight into alternatives to conflicts, business dilemmas, personal dilemmas, crises and helping people out of self-imposed psychological binds.

2. If your role is managing people, then your enlightened perspective will reveal ways to

cultivate behaviors or inspire people to act. This will be more rewarding than lording your authority over them (see Matthew 20:24-28).

3. Before presenting your enlightened perspectives to others, walk through the following:

Step 3.1: Identify all of the challenging or problematic consequences that might follow if the perspective were adopted.

Step 3.2: Consider the beneficial consequences that it affords.

Step 3.3: Instead of creating a final presentation, involve others in researching and discerning the necessary mitigations, incentives, disincentives, or reinforcements to include in a final presentation. This gives more people a personal investment in the outcome.

4. Use your enlightened perspective to help people see the false promises in perfectionism, binary thinking, and fear of making wrong decisions. You can help the stubborn to see that there is no real security or assurance in holding on to a particular viewpoint. Or you can reveal to the unhealthy competitive person what they are actually losing in their quest to win.

5. Use your enlightened perspective to depolarize people by involving them in activities that they commonly care about, along with people who are different. Often, the fellowship that ideology will not permit, shared experiences will. Experiencing things together can help each person to see the human being in the other, and possibly build a sense of community. (Note: Do not insist that people abandon their position, but aid them in refining their position, or to discover other possibilities)

6. Your enlightened perspective can make the case for an aloof person to get to know others through relationship. This can move them from the polar position where privilege has lulled them to

   • Prefer exclusivity over inclusion.

   • Quickly stand on moral law, rather than examining themselves introspectively, or seeking to understand others.

   • Not get their hands dirty, when they should, in order to aid a fellow human being.

   • Be callous rather than compassionate.

   • Be concerned with what is legal and moral at the expense of what is merciful and truthful. (Matthew 23:23-24; Romans 3:20; 7:7)

7. When there is polarization in the church, before one side becomes entrenched in their position, urge them to give space for the Holy Spirit to intervene. If they are willing, teach them the practice of contemplative prayer. Have everyone reconvene when ready. Later, determine if there are new viewpoints, refined existing viewpoints, changed viewpoints, or a change in the confidence level to one's viewpoint. Then work together to determine what movement can be made towards the other side's position.

By shining our light in the culture in this manner, our contemplative prayers can mitigate coercion, favoritism, or outcomes being influenced by egos. We will leave the people involved feeling a lot less disappointed and more supported. But more importantly, their movement from a polar end towards the center will be a non-threatening experience. Such weaponization of prayer will prevent divisiveness and the evils that follow. Then, the church can function more like the arms and legs of Christ.

About the Author

Jimmie D. Compton, Jr. and his wife Nancy have two children, two grandchildren, and two great-grandchildren. He received a master's degree in Pastoral Counseling from Ashland Theological Seminary and was a licensed therapist for 25 years. In addition to being the founding and Senior Pastor of Hope Bible Fellowship Church for over thirty years, Jimmie also provided counseling services for the Detroit Police Chaplain Corps, Detroit Rescue Mission Ministries, Eastwood Clinic, and New Way Christian Community Church. He has aided dozens of church clergy in biblical, counseling, and ministerial matters while authoring several books. Currently, he is Board Chairman of Citikidz Christian Sports Camp in Rector, PA. He loves running, walking, working out, and mentoring young men in his neighborhood.

Jimmie has been honored for completing an in-depth, six-year research project, *Early African Church History: From Jesus' Resurrection to the Rise of Islam*. This research has been developed into a two-year, online, self-paced curriculum at Hope Institute, the teaching and ministerial arm of Hope Bible Fellowship Church.

Related Companion Books

The book *Ball of Confusion in Galatia: Fighting a Religious, Nationalistic, and Ethnic Battle* highlights how Jesus and the Apostle Paul provided their faith communities with clarity during the cultural mishmash of ideological confusion during their day, while maintaining the ethos, pathos, and logos of the kingdom of God.

And, *In All Things, Love: Escaping the Church Schism Cycle* is a reflection of current events within the Christian church in America among Christian nationalist, Christian activist and carnal Christians. Infamous similarities to church schisms during the 4th through 5th century A.D. are cited, as well as cautions about the possibility of similar outcomes.

These, as well as other, books by Jimmie Compton are available on Amazon, Kindle and Audible. For bulk purchases, please email us at hbf.church@gmail.com with the title in the subject line. You can find details about his works at amazon.com/author/jimmiecompton